Pollinators

WASPS

Emma Bassier

DiscoverRoo
An Imprint of Pop!
popbooksonline.com

abdobooks.com

Published by Pop!, a division of ABDO, PO Box 398166, Minneapolis, Minnesota 55439.

Printed in the United States of America, North Mankato, Minnesota.

102019
012020

THIS BOOK CONTAINS RECYCLED MATERIALS

Cover Photo: iStockphoto

Interior Photos: iStockphoto, 1, 7 (flowers), 7 (wasps), 8, 12, 15 (bottom), 17, 20–21, 22 (bottom), 23 (bottom), 25, 27, 28; Shutterstock Images, 5, 9, 13, 31; Claude Nuridsany & Marie Perennou/Science Source, 6; Jerzy Gubernator/Science Source, 11; Michel Gunther/Science Source, 14–15; Larry West/Science Source, 18, 30; Robert and Jean Pollock/Science Source, 19; John M. Coffman/Science Source, 22 (top); Perennou Nuridsany/Science Source, 23 (top); Robert Wyatt/Alamy, 26; Nigel Cattlin/Science Source, 29

Editor: Connor Stratton

Series Designer: Jake Slavik

Library of Congress Control Number: 2019942506

Publisher's Cataloging-in-Publication Data

Names: Bassier, Emma, author.

Title: Wasps / by Emma Bassier

Description: Minneapolis, Minnesota : Pop!, 2020 | Series: Pollinators | Includes online resources and index.

Identifiers: ISBN 9781532165993 (lib. bdg.) | ISBN 9781532167317 (ebook)

Subjects: LCSH: Pollinators--Juvenile literature. | Wasps--Juvenile literature. | Hornets--Juvenile literature. | Pollination by insects--Juvenile literature. | Insects--Juvenile literature.

Classification: DDC 595.798--dc23

WELCOME TO DiscoverRoo!

Pop open this book and you'll find QR codes loaded with information, so you can learn even more!

Scan this code* and others like it while you read, or visit the website below to make this book pop!

popbooksonline.com/wasps

*Scanning QR codes requires a web-enabled smart device with a QR code reader app and a camera.

TABLE OF CONTENTS

CHAPTER 1
SPREADING POLLEN

The sun shines on a fig tree. A wasp lands on one of the figs. The wasp has **pollen** on its body from another fig. The wasp enters the fruit through a small hole. Tiny flowers grow inside

WATCH A VIDEO HERE!

A wasp crawls toward the opening of a fig.

the fig. The wasp mates and lays eggs on the flowers. Then it dies. The eggs hatch and grow into wasps. Some of the wasps get pollen on their bodies. Then they fly to other fig trees.

One reason wasps visit figs and other flowers is to mate and lay eggs. Another reason is for food. Many wasps sip flower **nectar**. Other wasps hunt insects that live on the flowers. Wasps sometimes carry the flowers' pollen on their bodies when they leave. That pollen often rubs off onto other flowers.

A cuckoo wasp becomes covered in pollen while feeding on nectar.

WASP POLLINATION

A wasp feeds on a flower's nectar. Pollen gets on the wasp's body. The wasp flies to another flower. Some of the first flower's pollen falls into the second flower.

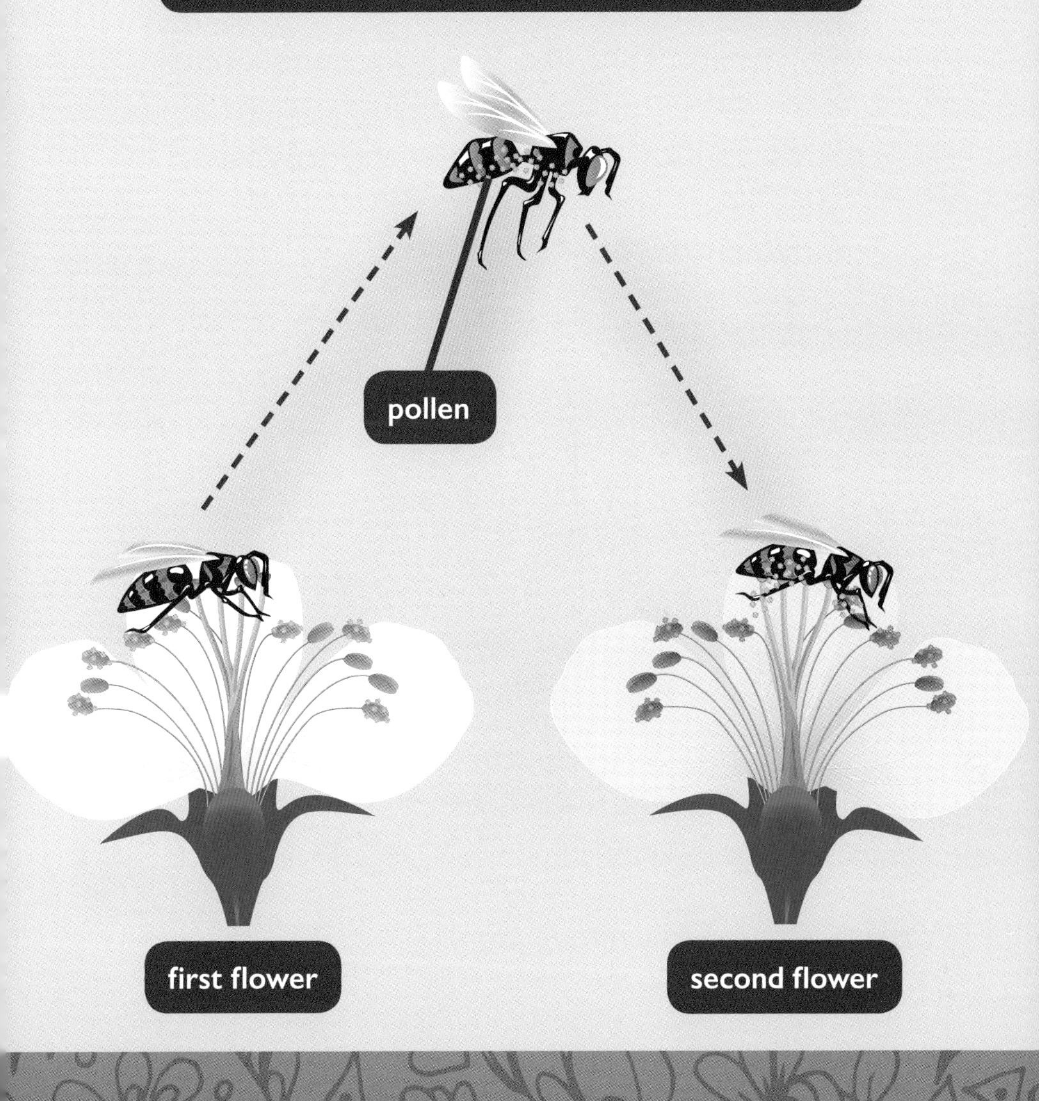

Insects such as wasps and bees are pollinators. They help plants survive. Many plants cannot create seeds unless their pollen spreads. Without seeds, new plants cannot grow. And without wasps, many plants could die.

Wasps have been pollinating plants for millions of years.

A wasp crawls inside the petals of a beardtongue flower.

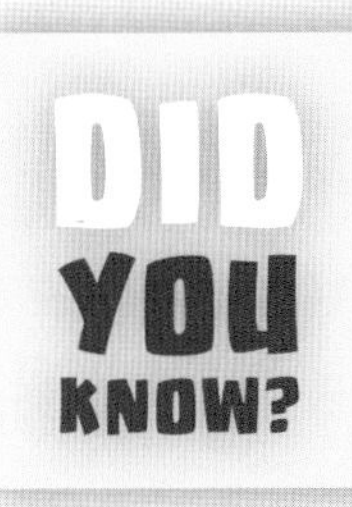

Beardtongue flowers depend on pollen wasps for pollination.

CHAPTER 2

WINGS AND STINGERS

More than 110,000 types of wasps exist on Earth. All these wasps fit into two main groups. These groups are predator wasps and parasite wasps. Wasps in both groups help pollinate plants.

LEARN MORE HERE!

Wasps can be any color. One type of predator wasp is blue and red.

Compared to bees, wasps have pointed bodies and slimmer waists.

Predator wasps usually hunt other insects. They feed these insects to their **larvae**. In contrast, parasite wasps lay

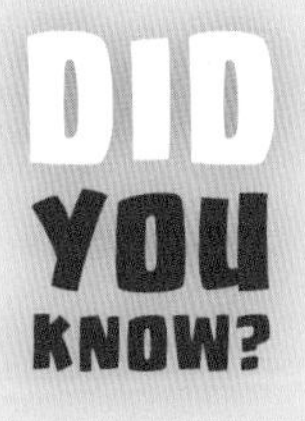

One type of parasite wasp can lay 200 eggs into a caterpillar cocoon at one time.

A pollen wasp visits a marigold flower.

their eggs on plants or in the bodies of other insects. Their larvae grow by eating the **host**.

POLLEN WASPS

One type of predator wasp is the **pollen** wasp. Unlike most predator wasps, pollen wasps do not hunt insects. Instead, they feed their larvae **nectar** and pollen. Adults tend to have long mouthparts. They can reach nectar in deep flowers. As a result, they are better pollinators than many other wasps.

Male wasps do not have stingers, but female wasps do.

Both groups of wasps can have stingers. Predator wasps use stingers to attack. Some even shoot poison out of their stingers. Parasite wasps do not attack with their stingers. They use stingers to lay eggs inside hosts.

Unlike bees, wasps can sting more than once.

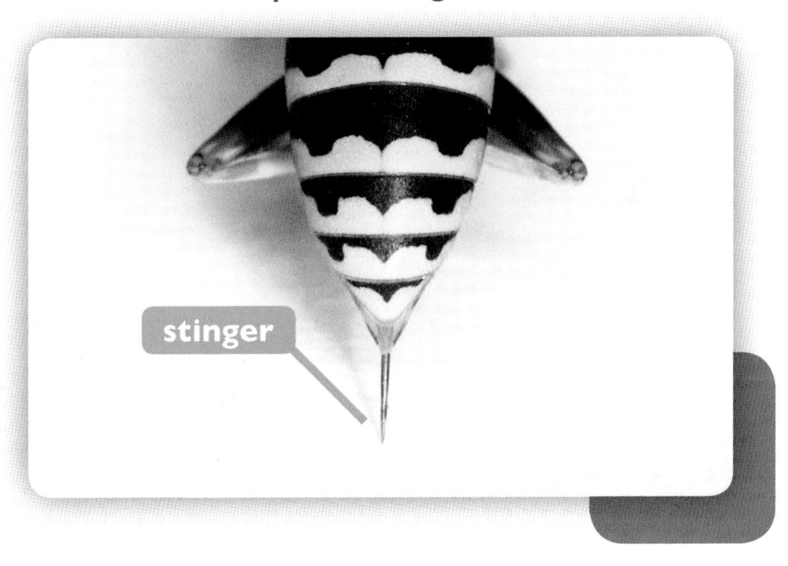

CHAPTER 3

WASP NESTS

Wasps live all around the world. Some live in mountains. Others live in forests or grasslands. In all these areas, wasps build nests. The nests help protect their **larvae**. Some wasps build nests

COMPLETE AN ACTIVITY HERE!

Wasp larvae grow inside the cells of a nest.

underground. Others build nests inside trees or walls.

Most wasps live alone. They are known as solitary wasps. These wasps build small nests. Many make nests underground. They dig tunnels in the dirt. Other solitary wasps build nests out of mud.

A solitary wasp digs a nest into the sand.

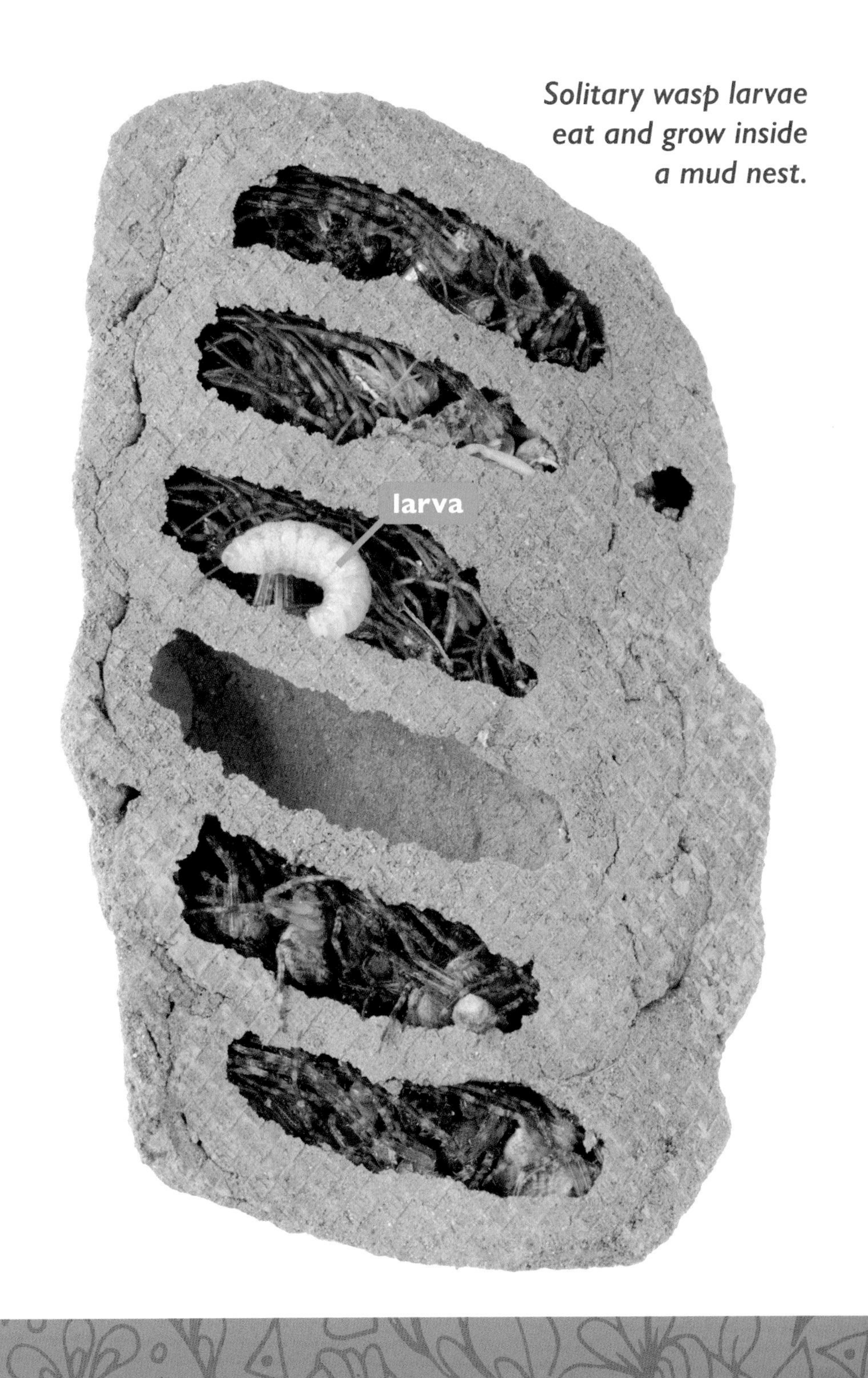

Solitary wasp larvae eat and grow inside a mud nest.

Some wasps live in **colonies**. They are known as social wasps. Every colony has one queen. She mates and lays eggs. Other wasps are workers. They find food and build the nest.

Although the queen begins building a colony's nest, workers take over to finish building.

Social wasp nests are often made of wood. Wasps scrape wood with their jaws. Then they chew the bits into **pulp**. They use that pulp to build the nest.

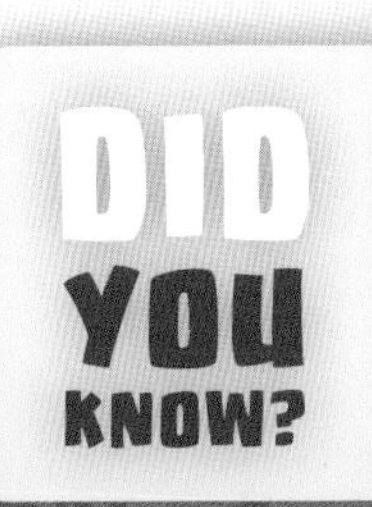

A colony of yellow jacket wasps can have up to 5,000 members.

LIFE CYCLE OF A WASP

A female wasp fills her nest with insects.

Each egg hatches into a larva. The larva eats the insects and grows.

She lays one egg in each part of the nest. Then she flies away.

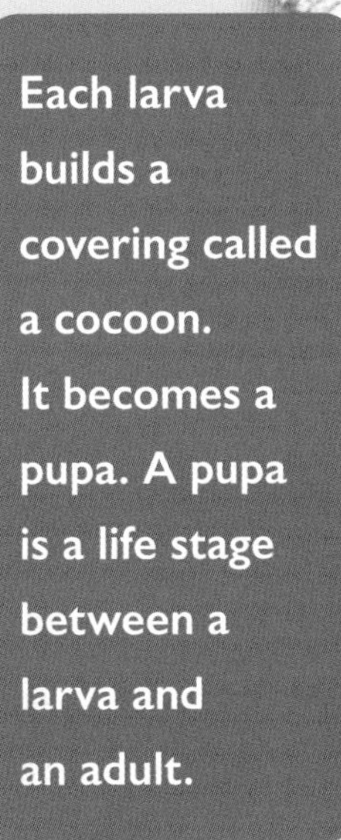

Each larva builds a covering called a cocoon. It becomes a pupa. A pupa is a life stage between a larva and an adult.

The pupa's body changes completely inside the cocoon. An adult wasp breaks out.

Many adult wasps live for only a few weeks. Some queens can live up to a year.

CHAPTER 4

HELPING OUT

Wasps play several important roles on Earth. For example, some wasps carry yeast in their stomachs. Yeast is a bacteria used to make bread. Yeast tends

LEARN MORE HERE!

The European wasp is one type of wasp that carries yeast with it through the winter.

to die off in winter. Wasps help it survive during cold months.

The King-in-his-carriage hammer orchid is pollinated by only one type of wasp.

In addition, wasps pollinate many flowers and crops. They are the main pollinators of many types of orchids and figs. The fig is an important fruit in many parts of the world.

Figs provide food for more than 1,200 types of animals.

Many farmers use wasps to protect their crops. Wasps eat large amounts of pests that harm crops. Without

Great black wasps can be found across North America.

Many wasps help protect grape plants.

wasps, these pests would be harder to control. By protecting wasps, humans can help a large number of plants as well.

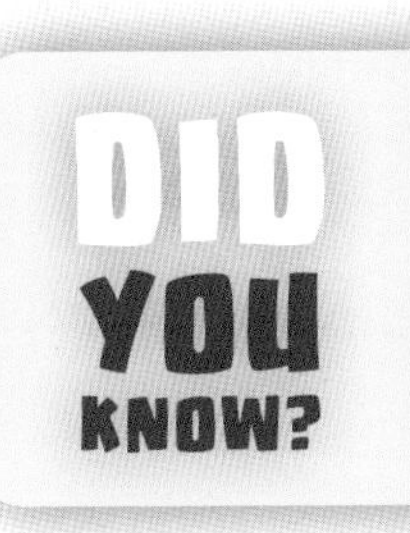

Black wasps help protect tomato plants. They help kill tiny insects called aphids.

MAKING CONNECTIONS

TEXT-TO-SELF

Have you seen wasps before? If so, what kinds?

If not, where might you find them?

TEXT-TO-TEXT

Have you read books about other insects?

What do they have in common with wasps?

How are they different?

TEXT-TO-WORLD

Some wasps live alone. Others live in colonies.

What might be helpful about living alone?

What might be helpful about living in colonies?

GLOSSARY

colony – a group of similar animals that live together in one place.

host – the living plant or animal that a parasite lives on or in.

larva – the first life stage of an insect, often looking like a worm.

nectar – a sweet, sugary liquid made by a plant.

pollen – fine, dust-like stuff that flowers create and use to reproduce.

pulp – soft or mashed up material, often wood.

INDEX

ONLINE RESOURCES

popbooksonline.com

Scan this code* and others like it while you read, or visit the website below to make this book pop!

popbooksonline.com/wasps

*Scanning QR codes requires a web-enabled smart device with a QR code reader app and a camera.